YOUR KNOWLEDGE HAS VALUE

- We will publish your bachelor's and master's thesis, essays and papers

- Your own eBook and book - sold worldwide in all relevant shops

- Earn money with each sale

Upload your text at www.GRIN.com and publish for free

Bibliographic information published by the German National Library:

The German National Library lists this publication in the National Bibliography; detailed bibliographic data are available on the Internet at http://dnb.dnb.de .

This book is copyright material and must not be copied, reproduced, transferred, distributed, leased, licensed or publicly performed or used in any way except as specifically permitted in writing by the publishers, as allowed under the terms and conditions under which it was purchased or as strictly permitted by applicable copyright law. Any unauthorized distribution or use of this text may be a direct infringement of the author s and publisher s rights and those responsible may be liable in law accordingly.

Imprint:

Copyright © 2017 GRIN Verlag
Print and binding: Books on Demand GmbH, Norderstedt Germany
ISBN: 9783668788190

This book at GRIN:

https://www.grin.com/document/438380

Jabar H. Yousif

Predictive Models for Photovoltaic Electricity Production in Hot Weather Conditions

GRIN Verlag

GRIN - Your knowledge has value

Since its foundation in 1998, GRIN has specialized in publishing academic texts by students, college teachers and other academics as e-book and printed book. The website www.grin.com is an ideal platform for presenting term papers, final papers, scientific essays, dissertations and specialist books.

Visit us on the internet:

http://www.grin.com/

http://www.facebook.com/grincom

http://www.twitter.com/grin_com

Article

Predictive Models for Photovoltaic Electricity Production in Hot Weather Conditions

Jabar H. Yousif [1], Hussein A. Kazem [2] and John Boland [3,*]

[1] Computing & Information Technology, Sohar University, P.O. Box 44, Sohar 311, Oman; jyousif@soharuni.edu.om
[2] Faculty of Engineering, Sohar University, P.O. Box 44, Sohar 311, Oman; h.kazem@soharuni.edu.om
[3] Centre for Industrial and Applied Mathematics, University of South Australia, Adelaide 5095, Australia
* Correspondence: john.boland@unisa.edu.au; Tel.: +61-8-830-23449; Fax: +61-8-830-25785

Academic Editor: Francesco Calise
Received: 23 May 2017; Accepted: 6 July 2017; Published: 11 July 2017

Abstract: The process of finding a correct forecast equation for photovoltaic electricity production from renewable sources is an important matter, since knowing the factors affecting the increase in the proportion of renewable energy production and reducing the cost of the product has economic and scientific benefits. This paper proposes a mathematical model for forecasting energy production in photovoltaic (PV) panels based on a self-organizing feature map (SOFM) model. The proposed model is compared with other models, including the multi-layer perceptron (MLP) and support vector machine (SVM) models. Moreover, a mathematical model based on a polynomial function for fitting the desired output is proposed. Different practical measurement methods are used to validate the findings of the proposed neural and mathematical models such as mean square error (MSE), mean absolute error (MAE), correlation (R), and coefficient of determination (R^2). The proposed SOFM model achieved a final MSE of 0.0007 in the training phase and 0.0005 in the cross-validation phase. In contrast, the SVM model resulted in a small MSE value equal to 0.0058, while the MLP model achieved a final MSE of 0.026 with a correlation coefficient of 0.9989, which indicates a strong relationship between input and output variables. The proposed SOFM model closely fits the desired results based on the R^2 value, which is equal to 0.9555. Finally, the comparison results of MAE for the three models show that the SOFM model achieved a best result of 0.36156, whereas the SVM and MLP models yielded 4.53761 and 3.63927, respectively. A small MAE value indicates that the output of the SOFM model closely fits the actual results and predicts the desired output.

Keywords: solar electricity prediction; artificial neural networks; photovoltaic; machine learning; self-organizing feature map (SOFM)

---

1. Introduction

The primary source of energy used to generate electricity is fossil fuel, which is a non-renewable energy source (RES) since it will run out in the future. Therefore, it is important to explore renewable energy sources such as solar, geothermal, wind, etc., which are geographically site-specific. Solar energy is one of the best options for Gulf Cooperation Council (GCC) countries. Solar energy presents as both thermal and light components. The latter can be converted into electricity using solar cells or photovoltaic (PV) technology, which is proven to work, and has been used for a long time. An important reason for utilizing PV technology is to reduce its price, which has indeed occurred, especially in the last decade, as depicted in Figure 1. Global production has increased by a factor of 370 since 1992 [1] as shown in Figure 2.

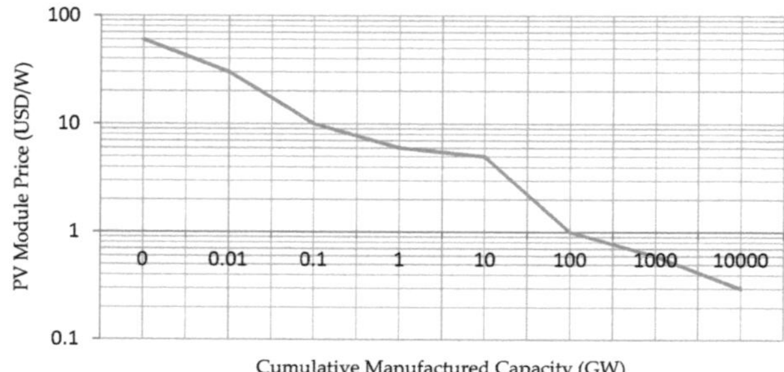

Figure 1. Projected economies of scale of photovoltaic (PV).

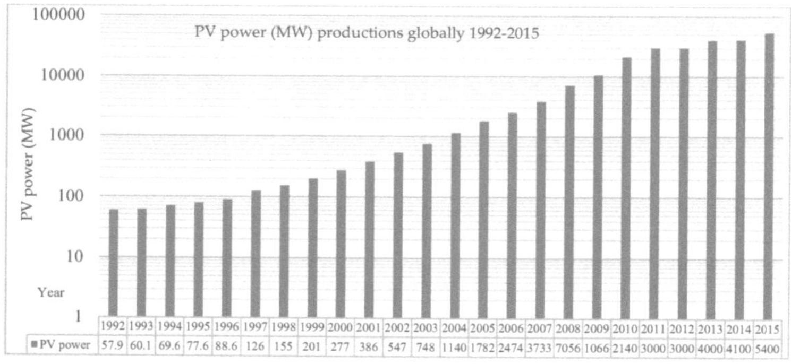

Year	1992	1993	1994	1995	1996	1997	1998	1999	2000	2001	2002	2003	2004	2005	2006	2007	2008	2009	2010	2011	2012	2013	2014	2015
PV power	57.9	60.1	69.6	77.6	88.6	126	155	201	277	386	547	748	1140	1782	2474	3733	7056	1066	2140	3000	3000	4000	4100	5400

Figure 2. Total PV power (MW) productions globally 1992–2015.

Electricity demand in Oman is expected to increase due to thriving industrial areas such as Sohar, Alrusail, and Alduqum. Furthermore, the population of the Sultanate of Oman has grown dramatically. This will increase the demand for electricity, as illustrated in Figure 3. In 2016, the maximum power demand reached 6000 MW, whereas the forecasted maximum power demand for 2018 is expected to reach 6800 MW [2]. Thus, more electrical power plants are needed. Energy sources must be diversified since approximately 98% of the fuel used to generate electricity in Oman is natural gas, while 2% is diesel fuel. Many studies have explored, evaluated, and investigated the feasibility of renewable energy systems and concluded that the priority for renewable energy generation in Oman is solar energy. Oman has 342 long sunny days (between 10 and 12 h) and large areas of free land in which PV systems can be installed [3,4]. Since most of the industrial areas and population (55%) are in the northern part of Oman and solar radiation has a high concentration, approaching 955 W/m^2 in summer, this region has become a target for PV investment.

PV output is affected by many factors such as solar radiation, temperature, humidity, dust, rain, and so on. Solar radiation and temperature are the main factors, so the power output of the PV system is a stochastic random process. Moreover, the PV power fluctuation due to the solar radiation and temperature fluctuations also affects the PV system capital and operation costs. The uncertainty of the PV system's power output is the major drawback of these systems. Therefore, this issue has encouraged researchers to propose models to forecast the PV output. These models will be used to forecast the long-term production of the PV systems, which is useful in terms of the technical and economic points of view.

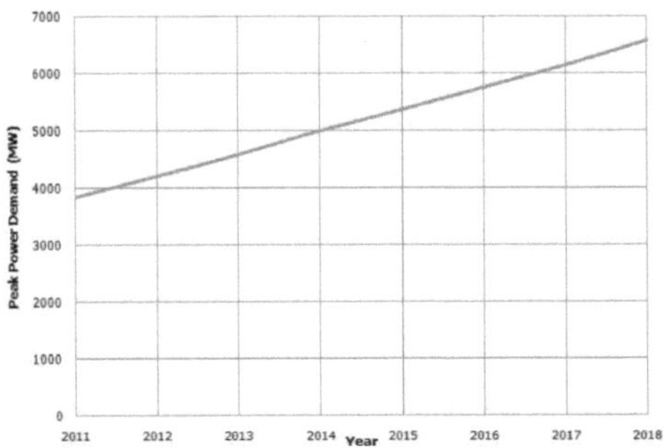

Figure 3. Oman peak power demand for 2011–2015 and projection until 2018.

Artificial neural networks (ANN) have become increasingly useful for system modelling and the optimization of results. ANN computation has increased our capacity to analyze and process data. ANN is a powerful modelling tool, which maps a complex input space into simple output space. In general, ANN applications tend towards the recognition, classification, prediction, generalization, and association of data. The network architecture determines the topology of the connections between neurons (input, hidden, and output) [5]. There are various topologies of ANN, such as feed forward neural networks and recurrent neural networks. In addition, ANN can be classified, based on learning techniques, into two categories: namely, supervised and unsupervised [6]. In this paper, ANN is used to predict the productivity of a photovoltaic system installed in Oman. Data were measured and recorded for one complete year in this study. The proposed self-organizing feature map (SOFM) model was tested and verified in terms of speed and accuracy. Accuracy is defined as the degree of a measured value that conforms to a known value or standard. In addition, the proposed model is discussed, analyzed, and compared with other models like the multi-layer perceptron (MLP) and support vector machine (SVM) models. Finally, the proposed model is compared with models in the literature for further verification.

2. Related Artificial Neural Networks-Based Work

Ogliari et al. [7] presented a hybrid intelligence computational evolutionary method for forecasting and analyzing the predictions of photovoltaic systems based on comparing different forecasting error techniques. The absolute hourly error, and daily absolute error, are the factors used to validate and analyze the errors of the forecasting models. They implement a case study for calculating error values during a clear sky day of radiation (theoretical, forecast, and actual).

Hernández et al. [8], developed an unsupervised 4×4 neurons self-organizing map (SOM) and the clustering k-means data processing system for analyzing the energy consumption patterns in industrial parks in Spain. The proposed system is validated based on real data and finding different behavior patterns that are meaningful and could work without any prior knowledge about the data.

Bracale et al. [9] proposed a short term probabilistic forecast model for predicting the hourly power of a photovoltaic system based on the probability density function. The model implements a Bayesian autoregressive time series for solar radiation based on critical variables like cloud cover and humidity. The proposed model is validated using the Monte Carlo simulation procedure based on a random sampling of the clearness index distribution.

In wind energy systems, Sheela and Deepa [10] had used a hybrid neural network model-based SOM and radial basis functions (RBF) to predict wind speed. Their experiments and proposed model results are close to the actual results with small percentage error. Several researchers are implementing supervised feed forward ANN techniques to forecast solar energy production [11–13].

Argiriou et al. [11] implemented a neural controller model based on feed forward back propagation for hydronic heating plants in buildings. The controller was used to forecast variables such as meteorological modules, ambient temperature, and solar irradiance. The proposed models are used on real-scale office buildings during real operating conditions. A comparison of operational results and the conventional controller is performed to evaluate the performance of the proposed numerical-based simulation models. The results of the experiments and numerical models showed that the percentage of energy saving is about 15% in North European weather conditions.

Khatib et al. [12] proposed a solar irradiation, system-based ANN, which uses data from 28 cities in Malaysia. The proposed neural network model is used to forecast the clearness index, which is used to establish a predictive global solar irradiation system for Malaysia. The predicted solar irradiation system has a mean absolute percentage error (MAPE) of 5.92%, and yielded 7.96% for the root mean squared error (RMSE) and 1.46% for the mean absolute error (MAE).

Dorvlo et al. [13] implemented an ANN technique to estimate the solar radiation-based clearness index. They used long-term data to design a hybrid neural, net system-based RBF and MLP for Oman. They claim that both the RBF and MLP models perform well. However, they suggested that RBF models are more efficient since they require less computing power. The RBF is trained with data obtained from different meteorological stations located in Sohar, Seeb, Masirah, Sur, Salalah, Sohar, and Fahud.

A number of works have involved implementing the feedback neural network, such as the recurrent neural network, and unsupervised learning or competition learning to predict renewable energy systems. Thus, ANNs have been widely implemented in renewable energy power systems. Moreover, several neural network techniques have been used to design and implement different phases of renewable energy power systems based on problem requirements and their characteristics [14–37].

Kalogirou and Bojic [14] implemented an artificial neural network model for predicting the energy consumption of a passive solar building of one room and a roof. A multilayer with a recurrent, architecture-based, back-propagation learning algorithm was implemented. The dynamic thermal building model was performed and tested for two seasons: winter and summer. The thickness of walls in simulated buildings varied from 15 cm to 60 cm. The proposed model was much faster than the dynamic simulation programs and achieved a good fit for actual data based on the value of the coefficient of determination (R^2 value), equal to 0.9985. Zhang and Chen [15] presented a predictive trend analysis model for energy consumption based on RBF network learning. The compression study with other feed-forward neural networks proves that the RBF model is faster and has a strong approximation function. The particle swarm optimization (PSO) algorithm is used for optimizing the performance of predicting energy consumption. Zhou et al. [16] proposed an artificial neural network model for estimating solar irradiation in Chinet, Beijing. The results of the proposed model showed that it can be used to evaluate the solar potential. Mohandes et al. [17] introduced the multilayer perceptron neural network for estimating global solar radiation in Saudi Arabia. The data was collected using 41 stations spread all over the country since 1971. These data were divided into training data sets (from 31 stations) and the other data (from 10 stations) were used as unseen data test set. The result of the proposed model proved the viability of the neural approach for modeling the solar radiation. The neural model obtained a MAPE equal to 4.49. Also, Adnan et al. [18] implemented a multilayer perceptron neural network for Turkey and it accurately predicted the actual data based on the value of coefficient of determination (R^2), which is equal to 0.958, with a MAPE equal to 6.7. Sozen et al. [19] proposed the use of the ANN for evaluating the solar energy potential in Turkey. They used different supervised learning algorithms like a scaled conjugate gradient (SCG), Levenberg-Marquardt (LM), and Polak-Ribiere conjugate gradient (CGP). The data for this paper is collected from 17 stations (namely cities) spread over Turkey for a period of 3 years (2000–2002). This model achieved a MAPE equal to 2.41, and the coefficient of determination (R^2) value was 99.99. Elminir [20] developed an ANN model based on the Levenberg optimization function for predicting the insolation data for different spectral bands of the Helwan monitoring station in Egypt. The proposed model obtained accuracies of 95%, 93%, and 96% for infrared, ultraviolet, and global insolation, respectively. Also, the model was tested based on infrared, ultraviolet, and global insolation using data for about one year for Aswan station, and it obtained accuracies of 95%, 91%, and 92%, respectively. Also, Elminir et al. [21] implemented a multilayer Perceptron Neural Network Neural Network model to predict diffuse fraction (KD) on an hourly and daily scale in Egypt. Hence, an effort is underway to implement the ANN model regarding a first order polynomial relating KD with the clearness index (KT) and sunshine fraction ($S/S0$). The ANN model showed that the model is suitable for predicting the diffuse fraction in hourly and daily scales, in comparison with the regression models. Rehman and Mohandes [22] utilized three feedforward neural network models for estimating global solar radiation (GSR) in the city of Abha in Saudi Arabia. The data were collected for 1462 days during 1998–2002, including factors like air temperature and relative humidity. The proposed model was tested using different combinations of input/output. The first model achieved a mean squared error (MSE) equal to 0.00028 and a MAPE equal to 10.3. The second model obtained MSE equal to 0.0052 and a MAPE equal to 11.8%. The last model obtained MSE equal to 0.00003 and a MAPE equal to 4.9. The testing of the proposed model proved that it was capable of estimating global solar radiation from temperature and relative humidity accurately. The study in [23] deploys a multilayer feed forward (MLFF) neural network model for evaluating the monthly average daily global solar irradiation on a horizontal surface in Uganda. The input factors of the neural model are sunshine interval, temperature, cloud cover, and location. The proposed model fits the actual data accurately, the correlation coefficient (R) is 0.974, the value of the mean bias error (MBE) is 0.059 MJ/m^2, and the value of RMSE is 0.385 MJ/m^2. The predictive results are compared and tested with other works in the literature. Mehmet et al. [24] applied a multilayered neural network model based on a resilient backpropagation learning algorithm for predicting the mean monthly wind speed of any target station in Turkey. The data was collected for a period between 1992 and 2001 by the Turkish State Meteorological Service (TSMS). The results demonstrated the proposed neural model could forecast the actual data accurately. It achieved a maximum value of mean absolute percentage error equal to 14.13% for the Antakya meteorological station and it got a best value in the Mersin meteorological station equal to 4.49%. Also, the model achieved a maximum value of the correlation coefficient (R) equal to 0.97 and a minimum value equal to 0.67. Bosch et al. [25] developed an ANN model to determine the daily global irradiation at stations located in a complex terrain. The collected data sets consisted of 3 years' data of daily global radiation from 12 different stations located in the north face of the Sierra Nevada National Park in the South East of Spain. The output of the neural model was tested and compared with the actual data, which proved that the neural model was very accurate. It showed that the proposed model was an efficient and easy methodology for calculating solar radiation levels, with great results like the RMSE of 6.0% and a MBE of 0.2%.

Fadare [26] developed a supervised MLFF neural network model based on the back-propagation learning method for forecasting the production of solar energy potential in Nigeria. The data sets used in this work were collected from 195 cities in Nigeria from the NASA geo-satellite database over a period of 10 years (1983–1993). The input consisted of 7 variables and their locations (latitude, longitude, and altitude), month, average of sunshine duration, temperature, and relative humidity. The solar radiation intensity is the output variable of the model. The monthly mean of solar radiation potential in northern and southern regions ranges from 7.01 to 5.62 and 5.43 to 3.54 kWh/m^2 day. The proposed model implemented different learning algorithms like the SCG and LM. Also, three different numbers of hidden layers (5, 10, and 15) were used, and the models achieved correlation coefficients (R) equal to 0.978, 0.971, and 0.956, correspondingly. Jiang [27] presented an artificial neural network model for estimating the solar irradiation in China. The collected data includes locations over all China for the period 1995–2004. The proposed model is tested and compared with measured values using mean percentage error (MPE), MBE, and RMSE. It achieved a value of 1.55, -0.04, and 0.746, respectively, with an accuracy of 94.81%.

3. Methodology

In this work, 24 PV modules were installed at Sohar University in Oman. The latitude and longitude of Sohar-Oman, the second large city, is 24 20 N, 56 40 E. The Sohar zone has a very good solar energy potential and, therefore, any PV system investment in this zone is expected to be very feasible. Electrical parameters such as voltage, current, and power were recorded and monitored for a period of about one year. The PV module rating is 140 W, 13.9% efficiency, 7.91 A maximum current, 17.7 V maximum voltage, 8.68 A short circuit current, and 22.1 V open circuit voltage. Figure 4 illustrates the configurations of the PV systems. Electrical parameters of the grid connected PV system (power, current, and voltage) were recorded and monitored.

Figure 4. The PV system.

Global solar radiation was measured between July 2013 and August 2014 using a BF5 Sunshine Sensor (Version 1, Cambridge, England). The sensors generate the data under different conditions, and some of these data are out of range or cannot be used directly. Therefore, proper pre-processing actions and data cleaning must be done before using these data. The calibration accuracy and sensitivity were ±0.12% and 1 mV/0.5 W m^{-2}, respectively, and the range of irradiance was 0–1250 W m^{-2}. The operating temperature was −20–+70 °C and an RHT2 temperature sensor (Version 1, Cambridge, England) was used. The data was downloaded using a DL2e Data Logger and specific software (Version 5, Cambridge, England). The average temperature in most Gulf countries is symmetric, as illustrated in Figure 5. Therefore, we can say the results of this work can be suitable for implementation in the whole Gulf region, with adjustable conditions.

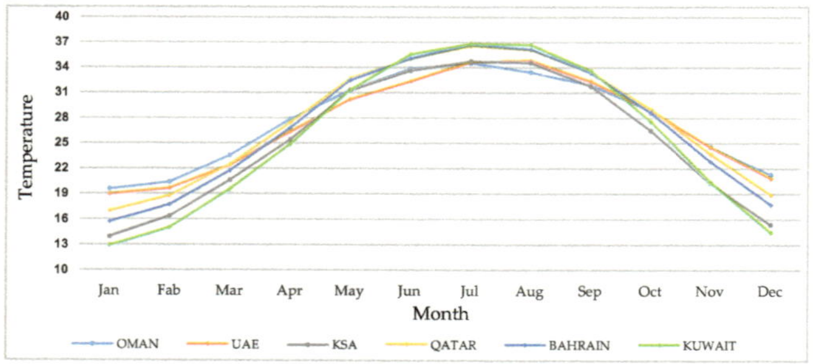

Figure 5. Average monthly temperature for the Gulf Cooperation Council (GCC) from 1901 to 2015 (source: worldbank.org).

4. Self-Organizing Feature Map Architecture and Design

SOFM was used to transfer multidimensional data to lower dimensional spaces. The differences between the input vector and the vectors of other neurons (D_{ij}) are calculated as in Equation (1):

$$D_{ij} = \left| X^l - W_{ij} \right| = \sqrt{(x_1 - w_{ij1})^2 + \dots\dots + (x_n - w_{ijn})^2} \tag{1}$$

The best matching unit (BMU) is the winning node whose weight vector is the most similar to the input vector.

$D(k1, k2) = \mathrm{min}i,j\ (D_{i,j})$, where $k1$ and $k2$ are the indexes of the winner-neuron. The weights of the winner node and its neighbor neurons are then adjusted. The neighborhood function is used to determine the neighborhood of a neuron as follows:

$$h(\rho, t) = \exp(\rho^2 / 2\sigma^2(t))$$

where ρ is the distance to the winner-neuron, which is computed as in Equation (2):

$$\rho = \sqrt{(k_1 - i)^2 + (k_2 - j)^2} \tag{2}$$

hat $h(\rho, t)$ or the French hat $h(\rho)$ as in Equation (3):

$$h(\rho, t) = \exp(-\rho^2 / \sigma^2(t)) \left[1 - \frac{2}{\sigma^2(t)} \rho^2 \right] \tag{3}$$

Thus weights of all the neurons are updated as in Equation (4):

$$W_{ij}(t + 1) = W_{ij}(t) + \alpha(t)h((\rho, t)(x^l(t) - W_{ij}(t)) \tag{4}$$

where $\alpha(t)$ is the learning rate. Finally, the weight vector of the winner-neuron or its adjacent is updated. The back-propagation learning algorithm is usually implemented in the MLP model to propagate the errors through the network. Error correction learning $e_i(n)$ is defined as in Equation (5):

$$e_i(n) = d_i(n) - y_i(n) \tag{5}$$

Gradient descent learning is performed to adapt each weight in the network, as in Equation (6):

$$w_{ij}(n + 1) = w_{ij}(n) + \eta \delta_i(n) + x_j(n) \tag{6}$$

where the local error $\delta_i(n)$ is computed from $e_i(n)$ at the output processing element (PE). The constant step size is η.

The SOFM is designed and implemented using the Neuro-Solutions software package. Figure 6 illustrates the design of SOFM, which has two PEs as input layers (solar radiation and ambient temperature), one output layer (the PV current), and one hidden layer. The data for this paper are generated from the twenty-four PV modules, which are shown in Figure 4 for a period of about one year (July 2013 until August 2014). The panel generates the data every hour. Therefore, we will use the average of each day as one data set. This work uses about 245 data sets (days), which splits into three categorized (40% as training data sets, 40% for the cross-validation data set, and 20% for testing data sets). The size and dimension of the unsupervised output space is 5×5. The neighborhood shape is significant in defining the organizational rules within the neural field. The final shape of the feature map greatly depends on the size of the initial and final radii of the neighborhoods. The final radius should have a small value, normally one or two PEs wide. Therefore, the initial radius is determined to be equal to 2 and the final radius is determined to be equal to 0. The learning rule and non-linearity selected for the hidden layer is momentum function with a step size of 1 and a momentum rate of 0.7. The TanhAxon is used as a transfer function in the four nearest neighbor PEs and output layers. The unsupervised learning rate is started with 0.01 and decays to 0.001. In order to enhance the behavior of the proposed SOFM model, several epochs (1000, 2000, 5000, and 10,000) are implemented. The training process is terminated when all the weights are less than or equal to a specific value.

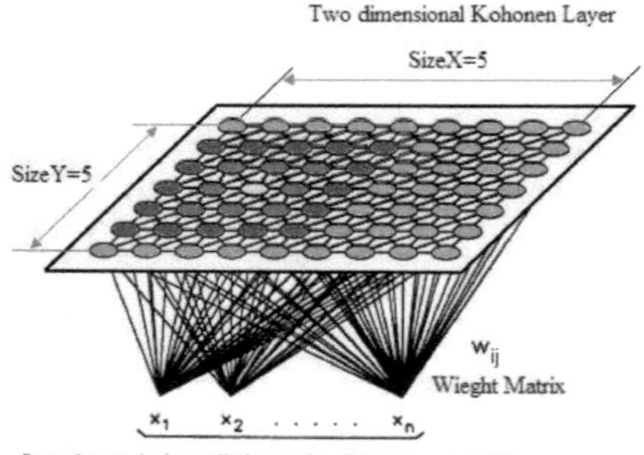

Figure 6. The self-organizing feature map (SOFM) architecture.

5. Self-Organizing Feature Map Architecture and Design

The NeuroSolutions package was applied for various methods to measure the performance of neural models and estimate the errors in the model output data. A well-known error measuring method is the *MSE*, which is the result of dividing the summation of the squared error (*SSE*) by the population *n*, as defined in Equations (7) and (8):

$$MSE = \frac{1}{n}SSE. \tag{7}$$

$$SSE = \sum_{i=1}^{n}(x_i - \overline{x})^2 \tag{8}$$

Here n is the number of observations, x_i is the value of the ith observation, and $\overline{x}$ is the mean value of all the observations.

Moreover, the correlation coefficient (r) is used to determine the linear dependence between two or more data sets. The values of r in the range of $(-1, 1)$. The correlation is computed as in Equation (9):

$$r = \frac{n\left(\sum_{i=1}^{n} x.y\right) - \left(\sum_{i=1}^{n} x\right)\left(\sum_{i=1}^{n} y\right)}{\sqrt[2]{\left[n\left(\sum_{i=1}^{n} x^2\right) - \left(\sum_{i=1}^{n} x\right)^2\right]\left[n\left(\sum_{i=1}^{n} y^2\right) - \left(\sum_{i=1}^{n} y\right)^2\right]}}. \tag{9}$$

where x is the first dataset $\{x_1, ..., x_n\}$ containing n values and y is the other dataset $\{y_1, ..., y_n\}$ containing n value.

5.1. Epoch Results

The experiments undertaken are implemented for different numbers of epochs (1000, 2000, 5000, and 10,000) in the training and testing of the SOFM model for the purpose of enhancing the performance of the proposed models. Figure 7 demonstrates the results of *MSE* for the SOFM-1000 model using 1000 epochs achieved 0.0005 in the training phase and 0.0006 in the cross-validation phase. The comparison of the proposed SOFM model and the desired output is illustrated in Figure 8.

It clearly indicates that the SOFM model predicts the actual results well. For the sake of testing the stability of the network, several different epochs were implemented. The graph for the final *MSE* of the proposed SOFM model using 2000 epochs achieved 0.0001 in the training phase and 0.0001 in the cross-validation phase, as illustrated in Figure 9. On the other hand, the desired output and actual network output using 2000 epochs are presented in Figures 10–12 illustrated the SOFM-5000, and SOFM-10000 models, which achieved *MSE*s of 0.00006, and 0.00003, respectively, in the training phase. Table 1 shows the results of minimum and final *MSE* for the SOFM network. Moreover, these models achieved *MSE*s of 0.0001, in the cross validation phase, as depicted in the Table 2. The experiments prove that increasing the number of epochs in the training and learning phases leads to a reduction in the value of *MSE*. In addition, the experiments prove that there is a strong relation between input and output variables based on the correlation factor of the SOFM-1000, SOFM-2000, SOFM-5000, and SOFM-10000 models, which are 0.9915, 0.9996, 0.9996, and 0.9997 respectively.

Table 1. Results of minimum and final mean square error (*MSE*) for the self-organizing feature map (SOFM-1000) network.

Best Networks	Training	Cross Validation
Epoch #	999	999
Minimum *MSE*	0.0007	0.0005
Final *MSE*	0.0005	0.0006

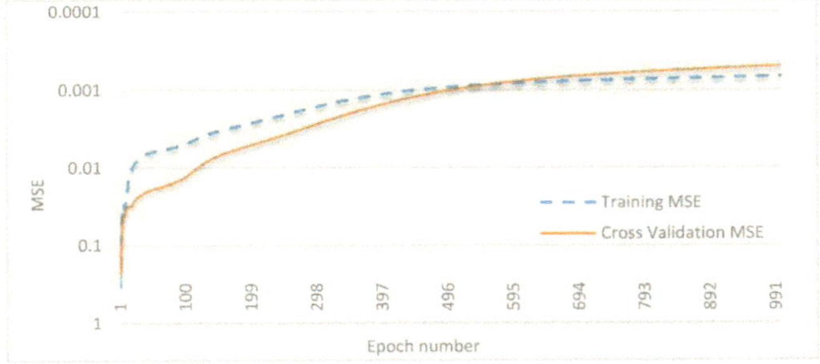

Figure 7. The mean square error (*MSE*) for the proposed SOFM model using 1000 epochs.

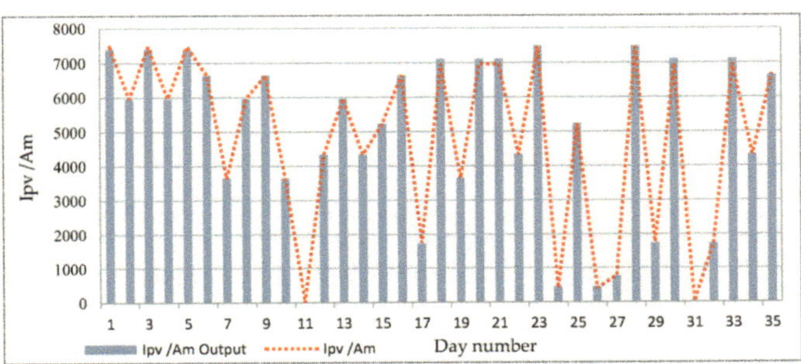

Figure 8. The testing results of the desired output and the proposed SOFM model output using 1000 epochs.

In order for the proposed SOFM model to be accurate and fit the actual results well, the coefficient of determination R^2 is used, which is defined as in Equation (10):

$$R^2 = 1 - \frac{\left(\sum_i (y_i - f_i)\right)^2}{\left(\sum_i (y_i - f\overline{y}_i)\right)^2}. \tag{10}$$

where y_i. is the observed value of the actual output, f_i is the predicted value, and $\overline{y}$. i is the arithmetic mean value of the observed targets. The preferred model is that which gets a coefficient determination R^2 value closer to 1. The prediction trend-line model for testing the SOFM-2000 model is calculated as in Equation (11):

$$y = 0.0884x^4 - 6.3087x^3 + 150.59x^2 - 1429.2x + 9474.5 \tag{11}$$

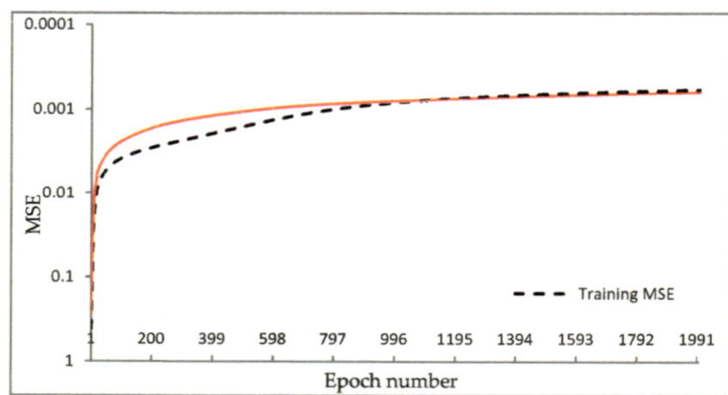

Figure 9. The *MSE* for the proposed SOFM model using 2000 epochs.

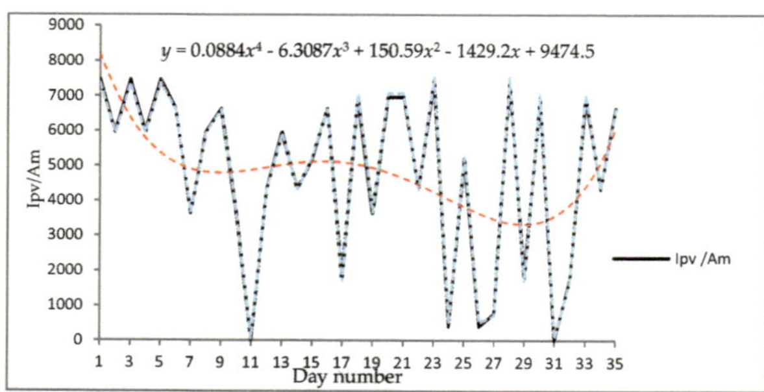

Figure 10. The testing results of the desired output and the forecasting SOFM output using 2000 epochs.

Figures 13 and 14 show the predicted trend-line models using a biquadratic polynomial function of degree 4 for SOFM-5000 and SOFM-10,000, respectively.

In general, the term accuracy is defined to mean the goodness of fit of measured results to the desired output. The SOFM models achieved an accuracy rate of 99%. MLP and SVM obtained accuracy rates of 92% and 62%, respectively.

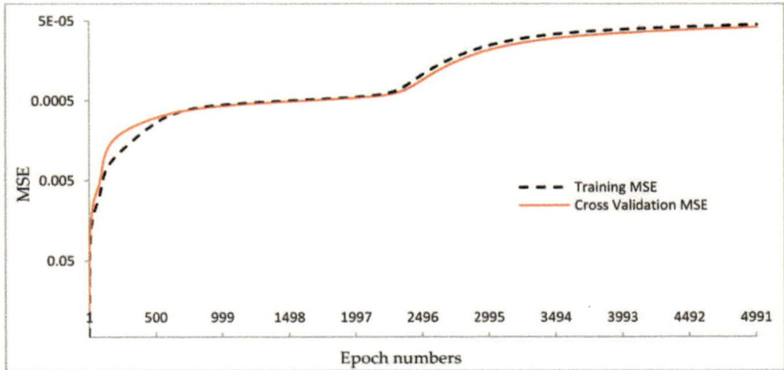

Figure 11. The testing of *MSE* for the proposed SOFM model using 5000 epochs.

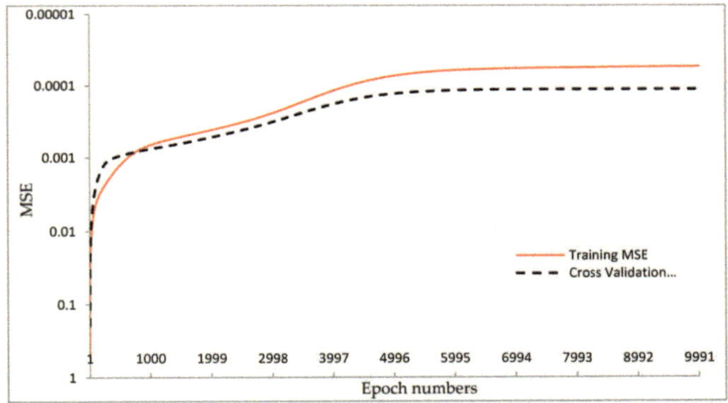

Figure 12. The testing of the *MSE* for the proposed SOFM model using 10,000 epochs.

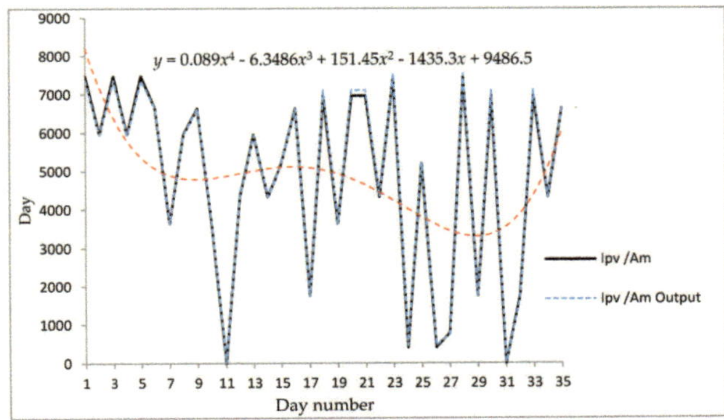

Figure 13. Testing the results of the desired data and the forecasting SOFM model using 5000 epochs.

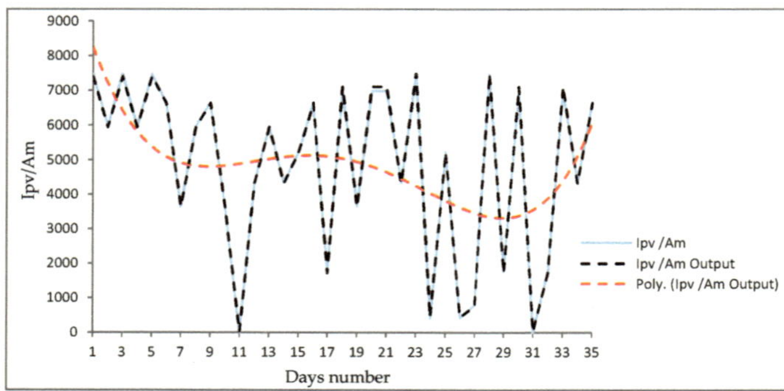

Figure 14. Testing the results of the desired data and the forecasting SOFM model using 10,000 epochs.

It is noted that the network results are non-linear functions, as shown in Figures 10, 13 and 14. Therefore, the equation of the trend line does not perfectly fit the results. For the purpose of ensuring that the proposed models achieve an excellent fitting curve for the desired results, the regression models were used to create a new fitting curve. The 2D-CurveTable software was used to create the new fitting model and it was then compared with the output of neural models. First, the data was cleaned and then used to draw the predicted data. The spline and Fourier smoothing estimation was performed in order to come up with excellent smoothing of data and good fitting. The experiment determines the confidence limit to 95% and then computes the fitting model, which is the best fit to the original data. The best model is computed as in Equation (12):

$$y = a + bx + cx^2 + dx^3 + ex^4 + fx^5 + gx^6 + hx^7 + ix^8 + jx^9 + kx^{10} \tag{12}$$

This polynomial model obtained a coefficient of determination R^2 value of 0.9554 and an F-value of 96.5827. The curve of the proposed model in Equation (12) is illustrated in Figure 15. The confidence limits are plotted in green and the forecasted interval values are plotted in red. The result of the regression model is plotted in blue using a dotted-line. The proposed model forecasts the electrical current production on a daily basis. It is clear that the polynomial model forecasts the future data for 60 days with a good fit. The original input data only does so for 35 days and the forecasting model is producing the current for 60 days.

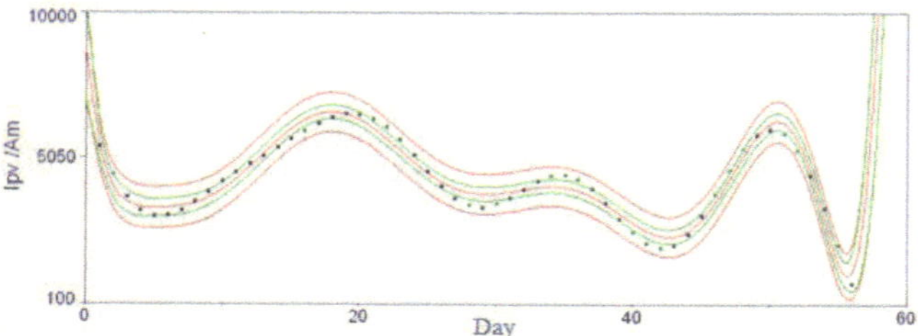

Figure 15. The result of the forecasting proposed model illustrated in the Equation (13).

Figure 16 shows the Comparison of the Desired Output and the Forecasted production of SOFM for 56 days. The actual amount of current produced (Ipv/Am) that is generated from the installed renewable power station PV appears in blue. And the forecasted production of current for 56 days is shown in red.

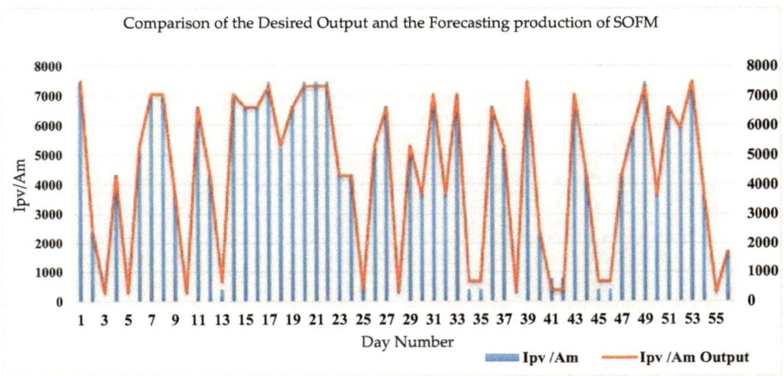

Figure 16. Comparison of the desired output and the forecasting production of SOFM.

5.2. Self-Organizing Feature Map, Support Vector Machine and Multi-Layer Perceptron Comparison

The comparison study covers all data sets (training, cross validation, and testing the results of the proposed models) to ensure that the measurement factors will apply for input and output data sets of neural models. The comparison was applied using several measuring criteria to verify the results of the proposed models, such as *MSE*, the *MAE* and coefficient of determination. Figure 17 demonstrates the comparison of the final *MSE* of the MLP, SVM, and SOFM models based on the training and cross-validation phases. This figure shows that the proposed SOFM model has achieved lower *MSE* values in comparison with the MLP and SVM models. The experiments prove that training the network multi-times with a varied number of epochs is necessary to enhance the results of neural models, as depicted in Table 2.

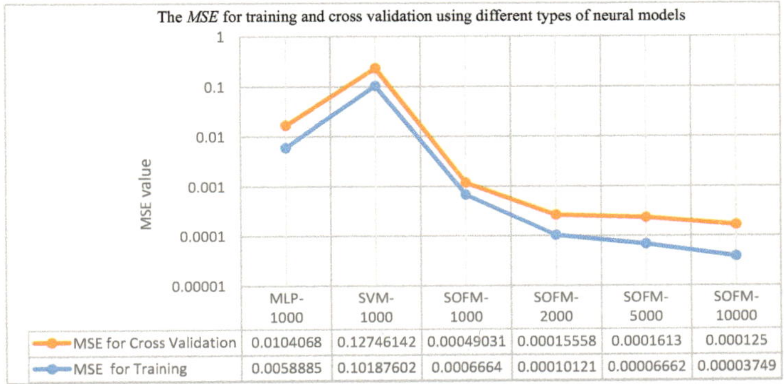

Figure 17. The comparison results of the final *MSE* for the multi-layer perceptron (MLP), support vector machine (SVM) and SOFM models.

Table 2. The comparison results of *MSE* in training and cross-validation, mean absolute error (*MAE*), *R*, and accuracy for multi-layer perceptron (MLP), support vector machine (SVM) and SOFM models.

Model No.	Neural Networks Type	*MSE* for Training	*MSE* for Cross Validation	*R*	*MAE*	Accuracy
1	MLP-1000	0.0058	0.010	0.9849	3.639	92%
2	SVM-1000	0.1018	0.1274	0.9820	4.537	61%
3	SOFM-1000	0.0005	0.0006	0.9915	0.456	93%
4	SOFM-2000	0.0001	0.0001	0.9996	0.531	99%
5	SOFM-5000	0.00006	0.0001	0.9996	0.435	99%
6	SOFM-10000	0.00003	0.0001	0.9997	0.3615	99%

MAE is appraising the proper fit of predicting values with desired values. It measures the forecast error in a time series analysis. In statistics, *MAE* is a factor used to measure the difference between the predicted value, the true value, and the expected range of errors that can be expected in the forecasting model. It is computed as in Equation (13):

$$MAE = \frac{1}{n} \sum_{i}^{n} \| f_i - y_i \| \tag{13}$$

where y_i is the observed value of the actual output and f_i is the predicted value.

The model that achieves a lower value of *MAE* is considered to be the best performing model. Figure 18 depicts the comparison of *MAE* values using different types of neural models. Obviously, the SOFM-10000 model achieved a lower *MAE*. Thus, it is considered as the best fitting model. Lastly, the SOFM models achieved a high accuracy of 99% in comparison with the MLP and SVM models, which obtained accuracy values of 92% and 61% respectively.

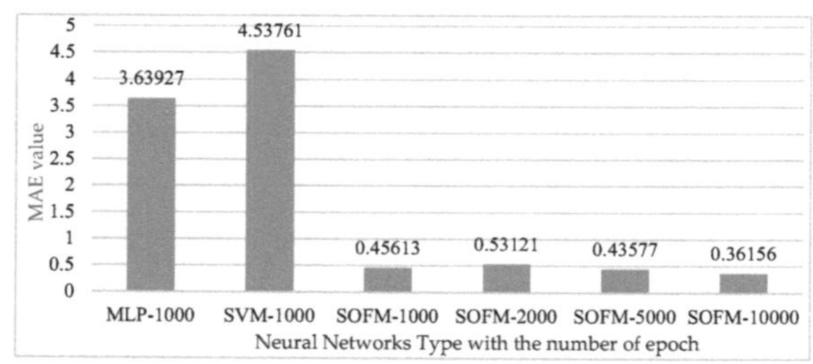

Figure 18. The comparison of MAE using different types of neural network models.

5.3. Comparison with the Relevant Models

Table 3 illustrates several representative examples of ANN techniques applied to the modelling or prediction of PV energy and solar radiation in the period 1998–2016. The results from the articles presented in Table 3 show that, in general, the errors associated with predictions (monthly, daily, hourly, and every minute) are between 3% and 14%. Moreover, MLP, SVM, and SOFM are the main techniques used to model or predict PV energy and solar radiation. However, it is noted that the MLP is the most commonly used technique and can be used with exogenous parameters or coupled with other predictors. The results shown in Table 3 illustrate the comparison of forecast accuracy. Factors such as MAPE, MBE, RMSE, *MAE*, *R*, R^2 and accuracy are used to verify the proposed models. In the *R* or R^2 and RMSE or *MSE* terms, the ANN model best fits the experimental data. The best results of the proposed neural model achieved an *MSE* value of 0.00003. Considering the same year (2016),

RMSE values of 10.09% and 5.60% were obtained in references [33,34], respectively. Furthermore, the proposed model (SOFM) has the highest accuracy of 99%.

Table 3. Comparison of production accuracy results.

Authors/Reference	Year	Location	Network Type	Model
Awada et al. [12]	2012	Malaysia	MLP	MBE: −1.12, RMSE: 8.57, MAPE: 7.29
Mohandes et al. [17]	1998	Saudi Arabia	MLP	MAPE: 4.49
Adnan et al. [18]	2005	Turkey	MLP	MAPE: 6.70. R value: 0.958
Elminir et al. [21]	2007	Egypt	MLP	SD: 9.00
Rehman and Mohandes [22]	2008	Saudi Arabia	MLP-FF1 MLP-FF2 MLP-FF3	MAPE: 10.3, MSE: 0.00028 MAPE: 11.8, MSE: 0.0052 MAPE: 4.49, MSE: 0.00004
Mubiru et al. [23]	2008	Uganda	MLP-FF	MBE: 0.059, RMSE: 0.385, R value: 0.971
Mehmet et al. [24]	2007	Turkey	MLP-Max. MLP-Min.	MAPE: 14.13%, R value: 0.97 MAPE: 4.49%, R value: 0.67
Bosch et al. [25]	2008	Spain	MLP-Max.	RMSE: 6%, MBE:0.2
Fadare [26]	2009	Nigeria	MLP-LM MLP-SCG	SD: 0.027, R value: 0.978 SD: 0.1104, R value: 0.961
Yingni J. [27]	2008	China	MLP	MBE: −0.04, RMSE: 0.746, MPE: 1.55, Accuracy 94.81%
Graditi et al. [33]	2016	Italy	MLP	RMSE: 10.09, R^2: 0.978
Huang et al. [34]	2016	USA	SVM	MAPE: 3.29, RMSE: 5.60, R value: 0.998
Current study	2017	Oman	SOF MML PSVM	MSE: 0.0001, 0.00007, 0.00003; MAE: 0.361, 3.369, 4.537; R value: 0.999, 0.984, 0.982; Accuracy: 99%, 92%, 61%

6. Sensitivity Analysis

Sensitivity analysis testing helps us to understand the relative importance among the inputs of the neural model and demonstrates how the output of the model varies in response to the variation of an input. Moreover, it helps to determine and eliminate an irrelevant input which reduces the data collection cost and can sometimes improve a network's performance. This test includes solar radiation and temperature. The sensitivity analysis of the SOFM model supports the view that solar radiation is the most influential variable on the output of the PV, as might be expected. Figure 19 depicts the sensitivity analysis results. It is well known that temperature and solar radiation play a central role in PV power generation and efficiency. Both power generation and efficiency have inverse and direct relationships with temperature and solar radiation, respectively. This result is important because, as the temperature in Oman is relatively high, which reduces the PV power productivity, the effect is low due to high solar radiation. Therefore, the large quantity of solar radiation in Oman is reflected positively in the PV performance, which reduces the importance of the reduction due to the increase in temperature.

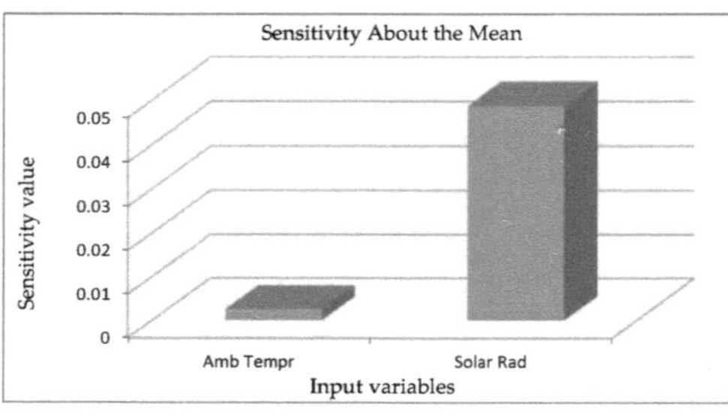

Figure 19. The sensitive analysis test for the SOFM network.

7. Conclusions

The main objective of this paper was to propose mathematical models to forecast the production of electrical current from photovoltaic panels using neural network techniques. Though the use of neural network methodologies to control and forecast current production is not new and many researchers have used and implemented them previously (15–33), it is worth noting that most of these applications are based on the use of supervised feed-forward neural networks. On the other hand, few studies have focused on recurrent neural networks and unsupervised models. The conclusions of this study contribute to the following:

(1) This paper is to propose a mathematical model to predict the production of electrical current from photovoltaic panels using the SOFM neural model, which is based on an unsupervised feedback neural network employed to improve the performance of the network. Moreover, this helps to reduce the error rate in predicting results generated by the neural model, when compared with the actual output. The data sets are divided into three sets (40% of the data sets for training the network, 20% of the data sets for determining the errors in training data set by using the cross-validation techniques, and 40% of the data sets for testing the output results of the network).

(2) The second contribution is performing the comparative study with the other two models (MLP and SVM) in order to ensure reliable results. The comparison study relied on several methods to measure the performance of the network, including the testing of the input and output data sets. These methods included *MSE*, *MAE*, the correlation factor, and the coefficient of determination. The proposed SOFM model achieves a final *MSE* of 0.0007 in the training phase and 0.0005 in the cross validation phase. This *MSE* is very small in comparison with other neural networks, such as that for the SVM model, which achieved an *MSE* of 0.0058. Moreover, the MLP model achieved an *MSE* of 0.026. The experiments prove that there is a strong relationship between input and output variables based on a correlation coefficient of 0.9989. In addition, the proposed SOFM model best fits the desired values based on the coefficient of determination R^2 value of 0.9555. For validating the results of the proposed model, *MAE* was used to confirm that the SOFM model closely fits the desired output, resulting in an *MAE* value of 0.361. SVM and MLP obtained *MAE* values of 4.537 and 3.639, respectively. Table 2 shows a comparison of *MSE* training, *MSE* cross-validation, *MAE*, *R*, and accuracy for the MLP, SVM, and proposed SOFM model.

(3) Another contribution involves implementing the sensitivity analysis to determine the degree of importance among the inputs of the neural model and demonstrating how the output of the model varies in response to a variation of an input. The sensitivity analysis of the SOFM model confirms that solar radiation is the most influential variable on the output of PV systems. Moreover, several epochs were implemented (1000, 2000, 5000, and 10,000), which helps to enhance the results of the proposed model. In addition, the accuracy of the proposed SOFM

model is compared with the accuracy of the SVM and MLP models. It was found that the accuracy of the SOFM model is 99%, while the accuracy of SVM and MLP are 62% and 91%, respectively.

(4) The last contribution of this paper involves implementing the regression techniques to generate a close-fitting model for the results of the SOFM model and the desired data. The best regression model was computed as in Equation (12), which is based on a polynomial equation. A coefficient of determination R^2 value of 0.9554 and an F-value of 96.582 were obtained. The curve of the proposed model of Equation (12) is illustrated in Figure 15. Finally, the proposed SOFM model predicts well and fits actual values based on a coefficient determination R^2 value of 0.9555. Actual values were gathered from solar irradiation and the production of the PV system of solar cells and the Photovoltaic research laboratory, which is installed at Sohar University.

Acknowledgments: The research leading to these results has received Research Project Grant Funding from the Research Council of the Sultanate of Oman, Research Grant Agreement No. ORG SU EI 11 010. The authors would like to acknowledge support from the Research Council of Oman.

Author Contributions: All the researchers were worked equally. Each of them helped in the writing of this paper and the design of the practical model and then proposed mathematical models.

Conflicts of Interest: The authors declare no conflict of interest.

Nomenclature

SOFM	Self-Organizing Feature Map
MLP	Multilayer Perceptron
MLFF	Multilayered feed-forward
SVM	Support Vector Machine
MBE	Mean Bias Error
RMSE	Root Mean Square Error
MSE	Mean Square Error
MAE	Mean Absolute Error
LM	Levenberg Marquardt
CGP	Polak-Ribiere Conjugate Gradient
SCG	Scaled conjugate gradient
R	Correlation
R^2	Coefficient of determination
GCC	Gulf Cooperation Council
PV	Photovoltaic
GSR	global solar radiation
ANN	Artificial neural networks
SOM	Self-Organizing Map
MAPE	mean absolute percentage error
RBF	Radial Basis Functions
BMU	Best Matching Unit
$k1, k2$	Indexes of the winner-neuron
ρ	Distance to the winner-neuron
$\alpha(t)$	Learning rate
$e_i(n)$	Error correction learning
$\delta_i(n)$	Local error
SSE	Summation of the squared error
x_i	Value of the ith observation
$\bar{x}$	Mean value of all the observations
x	First dataset $\{x_1, ..., x_n\}$
y	Other dataset $\{y_1, ..., y_n\}$ containing n value

References

1. Kazem, H.A.; Khatib, T. *Photovoltaic Power System Prospective in Oman, Technical and Economical Study*, 1st ed.; LAP LAMBERT Academic Publishing: Saarbrücken, Germany, 2013; ISBN 978-3659372957.
2. Authority for Electricity Regulation in Oman. *Study on Renewable Resources*; Final Report; Authority for Electricity Regulation in Oman: Muscat, Oman, 2008.
3. Altunkaynak, A.; Ozger, M. Comments on Temporal significant wave height estimation from wind speed by perceptron Kalman filtering. *Ocean Eng.* **2004**, *31*, 1245–1255.
4. Fakham, H.; Lu, D.; Francois, B. Power control design of a battery charger in a hybrid active PV generator for load-following applications. *IEEE Trans. Ind. Electron.* **2011**, *58*, 85–94.
5. Yousif, J.H. *Information Technology Development*; Academic Publishing: Saarbrücken, Germany, 2011; ISBN 9783844316704.
6. Kalogirou, S.A. Applications of artificial neural networks for energy systems. *Appl. Energy* **2000**, *67*, 17–35.
7. Ogliari, E.; Grimaccia, F.; Leva, S.; Mussetta, M. Hybrid predictive models for accurate forecasting in PV systems. *Energies* **2013**, *6*, 1918–1929.
8. Hernández, L.; Baladrón, C.; Aguiar, J.M.; Carro, B.; Sánchez-Esguevillas, A. Classification and clustering of electricity demand patterns in industrial parks. *Energies* **2012**, *5*, 5215–5228.
9. Bracale, A.; Caramia, P.; Carpinelli, G.; Di Fazio, A.R.; Ferruzzi, G. A Bayesian method for short-term probabilistic forecasting of photovoltaic generation in smart grid operation and control. *Energies* **2013**, *6*, 733–747.
10. Sheela, K.S.G.; Deepa, S.N. An efficient hybrid neural network model in renewable energy systems. In Proceedings of the 2012 IEEE International Conference on Advanced Communication Control and Computing Technologies (ICACCCT), Ramanathapuram, India, 23–25 August 2012.
11. Argiriou, A.A.; Bellas-Velidis, I.; Kummert, M.; André, P. A neural network controller for hydronic heating systems of solar buildings. *Neural Netw.* **2012**, *17*, 427–440.
12. Awada, A.; Pasupuleti, J.; Khatib, T.T.N.; Kazem, H.A. Modeling and characterization of a photovoltaic array based on actual performance using cascade-forward back propagation artificial neural network. *J. Sol. Energy Eng.* **2015**, *137*. [CrossRef]
13. Dorvlo, A.S.S.; Jervaseb, J.; Al-Lawatib, A. Solar radiation estimation using artificial neural networks. *Appl. Energy* **2015**, *71*, 307–319.
14. Kalogirou, S.A.; Bojic, M. Artificial neural networks for the prediction of the energy consumption of a passive solar building. *Energy* **2015**, *25*, 479–491.
15. Zhang, Y.; Chen, Q. Prediction of Building Energy Consumption Based on PSO-RBF Neural Network. In Proceedings of the 2014 IEEE International Conference on System Science and Engineering (ICSSE), Shanghai, China, 11–13 July 2014.
16. Zhou, J.; Wu, Y.Z.; Yan, G. Solar radiation estimation using artificial neural networks. *J. Sol. Energy* **2005**, *26*, 509–512.
17. Mohandes, M.; Rehman, S.; Halawani, T.O. Estimation of global solar radiation using artificial neural networks. *Renew. Energy* **1998**, *14*, 179–184.
18. Adnan, S.; Arcaklioglu, E.; Ozalp, M.; Caglar, N. Forecasting based on neural network approach of solar potential in Turkey. *Renew. Energy* **2005**, *30*, 1075–1090.
19. Sozen, A.; Arcaklioglu, E.; Ozalp, M.; Kanit, E.G. Use of artificial neural networks for mapping of solar potential in Turkey. *Appl. Energy* **2004**, *77*, 273–286.
20. Elminir, H.K. Estimation of solar radiation components incident on Helwan site using neural networks. *Sol. Energy* **2005**, *79*, 270–279.
21. Elminir, H.K.; Azzam, Y.A.; Younes, F.I. Prediction of hourly and daily diffuse fraction using neural network, as compared to linear regression models. *Energy* **2007**, *32*, 1513–1523.
22. Rehman, S.; Mohandes, M. Artificial neural network estimation of global solar radiation using air temperature and relative humidity. *Energy Policy* **2008**, *36*, 571–576.
23. Mubiru, J.; Banda, E.J.K.B. Estimation of monthly average daily global solar irradiation using artificial neural networks. *Sol. Energy* **2008**, *82*, 181–187.
24. Mehmet, B.; Sahin, B.; Yasar, A. Application of artificial neural networks for the wind speed prediction of target station using reference stations data. *Renew. Energy* **2007**, *32*, 2350–2360.

25. Bosch, J.L.; Lopez, G.; Batlles, F.J. Daily solar irradiation estimation over a mountainous area using artificial neural network. *Renew. Energy* **2008**, *33*, 1622–1628.

26. Fadare, D.A. Modeling of solar energy potential in Nigeria using an artificial neural network model. *Appl. Energy* **2009**, *86*, 1410–1422.

27. Jiang, Y. Prediction of monthly mean daily diffuse solar radiation using artificial neural networks and comparison with other empirical models. *Energy Policy* **2008**, *36*, 3833–3837.

28. Lam, J.C.; Wan, K.K.W.; Yang, L. Solar radiation modeling using ANNs for different climates in China. *Energy Convers. Manag.* **2008**, *49*, 1080–1090.

29. Kazem, H.A.; Khatib, T.; Sopian, K.; Elmenreich, W. Performance and feasibility assessment of a 1.4kW roof top grid-connected photovoltaic power system under desertic weather conditions. *Energy Build.* **2014**, *82*, 123–129.

30. Zhang, N.; Behera, K.P. Solar radiation prediction based on recurrent neural networks trained by Levenberg-Marquardt backpropagation learning algorithm. In Proceedings of the 2012 IEEE PES Innovative Smart Grid Technologies (ISGT), Washington, DC, USA, 16–20 January 2012.

31. Yona, A.; Senjyu, T.; Funabashi, T.; Mandal, P.; Kim, C. Decision Technique of Solar Radiation Prediction Applying Recurrent Neural Network for Short-Term Ahead Power Output of Photovoltaic System. *Smart Grid Renew. Energy* **2013**, *4*, 32–38.

32. Capizzi, G.; Napoli, C.; Bonanno, F. Innovative Second-Generation Wavelets Construction with Recurrent Neural Networks for Solar Radiation Forecasting, Neural Networks and Learning Systems. *IEEE Trans. Neural Netw. Learn. Syst.* **2012**, *23*, 1805–1815.

33. Graditi, G.; Ferlito, S.; Adinolfi, G. Comparison of Photovoltaic plant power production prediction methods using a large measured dataset. *Renew. Energy* **2016**, *90*, 513–519.

34. Huang, C.; Bensoussan, A.; Edesess, M.; Tsui, K.L. Improvement in artificial neural network-based estimation of grid connected photovoltaic power output. *Renew. Energy* **2016**, *97*, 838–848.

35. Kazem, H.A.; Yousif, J.H.; Chaichan, M.T. Modelling of Daily Solar Energy System Prediction using Support Vector Machine for Oman. *Int. J. Appl. Eng. Res.* **2016**, *11*, 10166–10172.

36. Yousif, J.H.; Kazem, H.A. Modeling of Daily Solar Energy System Prediction using Soft Computing Methods for Oman. *Res. J. Appl. Sci. Eng. Technol.* **2016**, *13*, 237–244.

37. Kazem, H.A.; Khatib, T.; Sopian, K. Sizing of a standalone photovoltaic/battery system at minimum cost for remote housing electrification in Sohar, Oman. *Energy Build.* **2013**, *6*, 108–115.

© 2017 by the authors. Licensee MDPI, Basel, Switzerland. This article is an open access article distributed under the terms and conditions of the Creative Commons Attribution (CC BY) license (http://creativecommons.org/licenses/by/4.0/).

YOUR KNOWLEDGE HAS VALUE

- We will publish your bachelor's and
  master's thesis, essays and papers

- Your own eBook and book -
  sold worldwide in all relevant shops

- Earn money with each sale

Upload your text at www.GRIN.com
and publish for free